Advanced Math Problems

For 4th and 5th

Grade Students

555 Practice Questions

www.555math.com

Tayyip Oral

Advanced Math Problems

For 4th and 5th

Grade Students

555 Practice Questions

www.555math.com

Published: January, 2024

ISBN: 9798874030261

Table of Contents

Preface

Advanced math book is a comprehensive resource designed to challenge and empower young mathematical minds. This book include 555 has difference type questions. Advanced math workbook for elementary school students that will improve their math fluency to the next level. This book helps students for math contests and strength the existing knowledge and prepare it for 6th grade math challenge. Advanced is designed to help improve a student's critical thinking, problem solving, and logical reasoning skills. This book will provide an overview of the different types of questions related to grade 4th and 5th. These questions can provide insight about areas in which students can enhance their reasoning skills, given.

Tayyip Oral

oral_tayyip@yahoo.com

TEST-1

1. 1+2+3…………………..+21=?

2. What number are missing?
   7, 14, 21, …, 42

3. It is currently 1:40 pm. What time will it be in 75 minutes?

4. What are the factors of 24.

5. What number added to 98 is equal to 124?

6. What are the common factors of 12 and 18?

7. Find the measure of an angle which is complementary to 27°.

8. Find the measure of an angle which is complementary of 14°?

9.   What is the measure of an angle that is supplementary to 154°?

10.   What is the measure of an angle that is supplementary to 127°?

11.   The sum of two integers is 44 and the difference is 4. What is the smaller of the two numbers.

12.   The sum of three consecutive even integers is 36. What is the value of the largest integer?

13.   One in integer is six times another. The sum of the integers is 56. What is the value of the largest number?

14.   The sum of the two numbers is 48 and one number are five times the other number. Find the great number square.?

15.   Seven times a number decreased by 6 is equal to 50. Find the number square.?

16. $3\frac{1}{4} \div 4\frac{2}{5} = ?$

TEST-2

1. Six times a number increased by 7 is equal to 37. Find the number.

2. Round of 789.7653 to the nearest hundredths?

3. Jack has 25 pennies. He found some more and now he has sixty-three. Which number sentence could be used to find how many pens he found?

4. There are seven apples in the basket. Juan cuts each of them into five pieces to serve the family. How many total pieces of apple are there?

5. Each pen costs is $1.20. How much does the entire pen cost, if there are 3.

6. Twenty-eight minutes ago, it was 8:54 pm. What time is now?

7. Calculate: 16.79+7.39 +4554+ 9.87

8. 98.76-14.65-25.65

9.   The ratio of the interior angles of a quadrilateral is 1:2:3:4. Find the measure of the largest angle.

10. $\angle A$ and $\angle B$ are supplementary angles. If $\angle A = 3x+20°$, $\angle B = 2x+40$. Find value of $3x$?

11.   Ahmet math quiz scores are 74, 84, and 78. What is his quiz average?

12.   If 6 tractors plough a field in 20 days, how many days does it take for 8 tractors to plough the same field?

13.   Jack family have three cars. The cars insurance cost is $156.72 per month. How much does it cost 2.5 years?

14.   140 of the total 300 passengers in an aircraft are female. What is the ratio of female passengers to male passengers?

15.   Jack's book is 288 pages. He read the 5/12 to the book. How many pages he did not read?

16. $9\dfrac{4}{6} \div 3\dfrac{7}{8}$

TEST-3

1.   How many sides does a hexagon have?

2.   Select the symbol to make this statement true 74+39_____ 89

3.   What is 2023+3+5+0+1 equal to?

4.   A number has seven ones, eight tens and six hundred. What is the number?

5.   What is the sum of the first six even whole number?

6.   Ahmet has 9 coins. His brother, Murat has twice as many coins as Ahmet. How many coins do they have altogether?

7.   How many times does the digit "4" appear from 3 to 53?

8.   How many triangles are there in the drawing?

9. If two sides of a triangle are 15 cm and 19 cm long. What are the largest possible values for the length of the third side?

10. If two side of a triangle are 17 cm and 21 cm long. What are the smallest possible values for the length of the third side?

11. In $\triangle ABC$, $\angle A = 65$, $\angle B = 77$. Order the sides of the triangle from least to greatest?

12. If the sum of 2/3 and 4/5 of a number is equal to 44, what is this number equal to?

13. Mario paint of the home wall Monday 1/7, Tuesday 2/3. Wednesday paint 1/6 part. What part did he finish wall paint?

14. What is the solution of 6x-4=34+4x?

15. 5/8 of students register to after school club. What percent of the students did not register club?

16. $16\dfrac{1}{4} \div 15\dfrac{1}{3}$

TEST-4

1. How is the number 975 read in expanded form?

2. 7 hundred, 6 tens and 9 ones are equivalent to which number?

3. Look at the number pattern below. Which number comes next?
    4, 8, 12, 16,….. ?

4. There are 4 cow, 3 lions and 5 eagles are at the zoo park. How many legs are there?

5. Ronaldo shaded some part in the circle.
    What fraction of the circle is shaded?

6. 7(7-7) + 7(7-6)+9×9 =?

7. Look at number 74. If the numeral 4 is replaced by 7, how would the number change?

8. 24+(24:4)×4+4

9. The ratio of two supplementary angles is 7:11. Find the great angle?

10. The ratio of the measures of the angles in a triangle is 2:6:7. Find the smallest angle?

11. The ratio of the measures of the angles of a triangle is 4:5:9. Find the greatest angle?

12. The ratio of the angles in a tringle is 1:5:9. What is the measures of the smallest angle?

13. Juan has $21. He spent $2.32 for pencil and $8.73 for math book. How much money does he have left?

14. Mario purchased 8 math practice books at $4.78 each. How much did he pay?

15. The sum of two consecutive integers is 31. What is the product of these numbers?

16. $2\frac{1}{2} + 3\frac{1}{3}$

TEST-5

1. 99+88+77=?

2. What digit shows hundreds in the number 32847?

3. 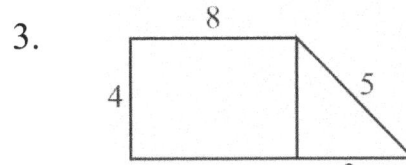 Perimeter=
   Area=

4. Solve: 5+125+235+345+465=?

5. Estimate: 44+79+77+96=?

6. Find the common multiply of 14 and 35.

7. Calculate: $(4+1/4)x(5-1/5)$

8. Today Ahmet added his age and his brother's age and he got 15. What will be the sum of their age after four years?

9. Triangle ABC and DEF are similar. AB= 9, BC= 6, and DE= 15, Find the EF?

10. 97.123-2.12-7.84

11.

Perimeter =

Area =

12. Triangle ABC and DEF are similar. The perimeter of smaller triangle ABC is 34 cm. The length of two corresponding sides on the triangles are 7 and 28. What is the perimeter of DEF?

13. What is the solution of $\frac{3}{5}m = 12$?

14. What is the solution of $\frac{2m}{3} - 6 = \frac{m}{3} + 16$?

15. If the sum of the ages of three siblings who are older than 5 years is 33 today, what was the sum of their ages 4 years ago?

16. $4\frac{3}{5} + 2\frac{1}{5}$

TEST-6

1. One pipe can fill a swimming pool in 6 hours, the second pipe can fill the same pool 10 hours. If they open together, how long hour fill pool?

2. AB=5 cm, AC=3 cm, BD=2 cm, DC=5 cm, AD= 6 cm. Find the sum of all triangle perimeters.

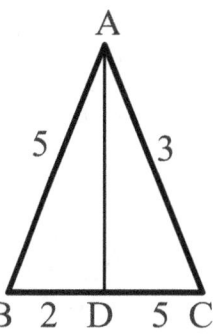

3. The product of 2 even numbers are always.

4. 2×1×55x0×99x77x88=?

5. What is the remainder when 17+17+17+17 is divided by 5

6. How many numbers are there in the sequence 4, 8, 12, 16,….,64, 68?

7. How many times does the digit "5" appear from 7 and 77.

8. The farm has 40 turkey and cow altogether. There are only 160 legs. Find the number of cows in the farm.

9. ABCD is rectangle. AB÷BC= 2:7 perimeter is 99 cm. Find the area of rectangle.

10. ABCD is a Square. The area of the square is 441 cm2. Find the perimeter of square.

11. ABCD is trapezoid. The area of trapezoid is 94 cm2. The height is 4 cm. Find the measures of bases.

12. A 3 gallon of milk weighs about 24.69 pounds. How many gallons of milk are there in 192.52 pounds?

13.

    Perimeter =

    Area =

14. A box contains 192 books. ¼ of the books are math. 1/6 of the books are geometry. How many of the books are geometry?

15. What is the solution of $\frac{2m+3}{5} = \frac{m-4}{6}$?

16. Compare: $\frac{23}{22}$ $\frac{11}{10}$

TEST-7

1. The New School soccer team has won 0.44 of its games this season. How can a new school express this decimal as a fraction?

2. Subtract: 2497-1642

3. Calculate; 24×11/8

4. Use the equation below to answer the question
   8×(2084) = □, which expressions could be used to correctly fill in the
   □

5. $\frac{1}{5} \div \frac{1}{5} \div \frac{1}{5} \div \frac{1}{5}$=?

6. Round 87.44872 to the hundredths place?

7. DC= 12cm, BC= 10 cm, LH=8 cm, HE= 4 cm.
   Find the shaded

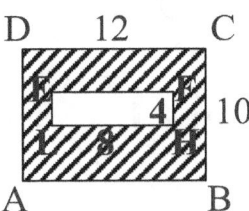

8. How many rectangles are there in the above figure?

9.　　The perimeter of a square is 48 cm. What are the 24% of area?

*　　The perimeter of rectangle is 180 cm and its length is 20 cm more its width

10.　　Find the 40% of area?

11.　　A carton of book is 19 cm long, 7 cm wide, and 21 cm tall. What is the volume of the carton?

12..　　The length of a rectangle is 4 cm more than the width. The perimeter is 36 cm. Find the rectangular area in square centimeters.

13.

Perimeter=
Area =

14.　　The sum of the digits of a number x is equal to 35. What is the remainder when x^2 is divided by 9?

15.　　The length of a rectangle is 5 cm more than the width. The perimeter is 50 cm. Find the area of a rectangle.

16.　　Set A {3, 7, 9, 11, 14,14}. Find Mean, mode, median and range

TEST-8

1. The product of all the factors of 12 is equal to ….

2. The letters in the word DALLAS were put in a box. What is the chance of getting letter A?

3. How many numbers are there in the sequence 12, 16, 20, ……………
    + 72, 76

4. Find the perimeter of ABCD rectangle.

    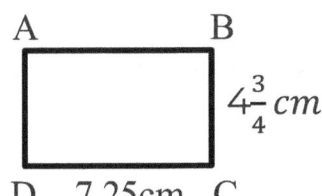

5. Which of the following is a factor of 68?

6. Calculate: $(99 \times \frac{12}{9}) \div 11$

7. Calculate: (99:3):3+3

8. The area of the square is 900 cm². Find the 27% perimeter of square.

9.   A rectangle has an area of 432 cm². If its length is 12 cm. Find its perimeter?

10.   The volume of a rectangular prism is 616 cm³. If the base of the prism is 11 cm by 8 cm. Find the height of the prism?

11.   Calculate: 9.88+8.77+7.66=?

      SET A = {2, 3, 7, 12, 8, 10, 10, 18, 26}

12.   Find the mode of the sat A.?

13.   Find range of the set A.?

14.   Find the median of seat A.?

15.   The area of Square is 81cm square. Find 30 % of the perimeter of the square?

16.   A rectangle swimming pool's length is x+6 m and width x+5 m. Find the perimeter of the swimming pool.?

TEST-9

Write least to greatest. (1-4)

1. $\frac{1}{3}$, $\frac{2}{4}$, $\frac{5}{8}$

   A   B   C

2. 36%,   0.32,   -0.64, $\frac{5}{8}$

   A   B   C   D

3. $\frac{4}{8}$, $\frac{2}{3}$, $\frac{1}{2}$, $\frac{4}{5}$

   A   B   C   D

4. 8%, $\frac{9}{100}$,   0.36,   0.07

   A   B   C   D

Find the rule.

5.

X	1	2	4
y	4	8	16

Rule: y =

6.

x	0	2	4	6
y	3	5	7	9

Rule: y =

7.

x	5	7	9	10
y	11	13	15	16

Rule: y =

8.

x	1	4	5
y	2	8	10

Rule: y=

9.

x	3	5
y	12	14

Rule: y=

10. The sum of four consecutive odd numbers is 56. What is the least of these numbers?

11. a, b, c and d being prime numbers and a<b<c<d, if axb=15, cxd=77, what is the sum of a+2b+3c+4d equal to?

12. 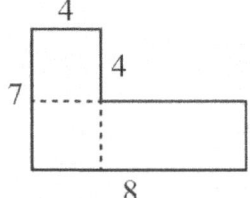   Area=

Perimeter=

13. What is the solution of $\frac{m}{6}+9=19$?

14. While numbering a 97 pages book, how many times is the numeral "6" used?

15. Calculate: (6-1/6)/(5-1/5)

16. One integer is three times another. Their product is 75. Find the great number.

TEST-10

Use the equation to complete each table.

1.

x- Input	1	4	7
y- Output			

$y=4x$

2.

x	1	3	5	9
y	3	9	15	?

$y=3x$

3)

x	0	5	7	12
y	4	9	11	?

$y=x+4$

Translating linear inequalities.

4. 7 is not more than x: _____

5. y is greater than 20: _____

6. 8 is not less than x: _____

7. x is not more than 9: _____

8. Solve: $\dfrac{a}{a+4} = 4 + \dfrac{3a}{a+4}$

9.   The quotient y and 4 is less than or equal to 10.

10.   What is the maximum sum of two numbers whose product is equal to 70?

11.   If the sum of two consecutive even numbers is equal to 34, what is the smaller number?

12.   If two consecutive odd numbers sum up to 44, what is the greater number?

13.   The sum of three consecutive even numbers are equal to 48. What is the greatest of these n?

14.   Three consecutive positive numbers sum up to 36. What is the middle of these numbers equal to?

15.   Decide whether each set of numbers can form a triangle 2cm, 4cm, 9 cm.

16.   A rectangle room measures 4x+4 and 3x+3 m. What is the perimeter and area of the room?

TEST-11

1. A rectangle prism volume 288 cm3, a length of 6 cm, a width of 3 cm. Find surface area of the prism?

2. Volume of prism is 640 cm^3, L= 10, W= 20 cm, height=?

3. Volume of a cube is 1000 cm^3. Find the base area.

4. Volume of a cube is 27 cm^3. Find the surface area.

5. Volume of a cube is 125 cm^3. Find the lateral area.

6. Five more than twice the number is 35. What is the number?

7. 2+4+6+................................36?

8. 1+3+5+....................................21?

9. Find all positive divisors of 88.

10. Seven times a number a is equal to 51 more than four times itself. What does the square of the number equal?

11. One of the two numbers whose sum is equal to 60 is 5 times the other one. What is the product of these two numbers?

12. One of the two numbers whose product is equal to 150 is 6 times the other one. What is the greater number equal to?

13. Calculate: (1/2+1/3+1/4+1/5)

14. Calculate: 9.12×21.9

15. 999×3-888×3=?

16. Jack designed a flag that measured 12 inch and 7 inches. Find the perimeter and area of the flag.

TEST-12

1.   0.7+7.7+7.77=?

2.   8.972+89.72+897.2=?

3. $1/3 \div 1/3 \div 1/3 \div 1/3$ =?

4.   (3 - 1/3) + (3 + 1/3) =?

5.   Simplify $6(36 \div 6) \div 6 + 6$=?

6.   Round of 278.6745 to the nearest hundredth

7.   97/5= a b/c,
     a + b + c=?

8.   What are the factors of 28.

9.   What are the common factors of 16 and 24

10.   Find the greatest common factor of the 6 and 18.

11.   What is the least common multiply of 2, 3 and 5?

12.   What is 20% of 40?

13.   Jack's ages is x and Mehmet is y years old. What will the sum of their ages be in t years?

14.   The sum of Jack's and Mehmet's ages is 2x. Which of the following gives the sum of their ages t years ago?

15.   The ratio of Jack's age 6 years ago to his age in 4 years is 3/8. How old is Jack now?

16.   A 24-year-old mother has a 3-year-old daughter. After how many years will the mother's age be 4 times that of her daughter?

TEST-13

1. 30 is 30% of what number?

2. Find the percent decrease from 120 to 80.

3. What is the arithmetic meaning of 2, 4, 6, 24, and 98.

4. Set A {2,4,6, 12, 12, 8, 10, 24} Find range, mean; median and mode.

5. Find the least common multiply of 8 and 18.

6. How many students speak three languages.

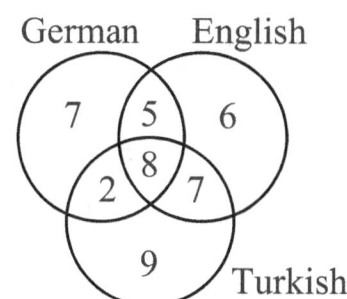

7. ABCD is a rectangle. AB=10cm, BC= 6cm. Find the ratio area to perimeter of rectangle.

8. The sum of three consecutive integer is 51. What is the largest number?

9. Two integers total 42. One integer is 28 larger than the other. What are the two integers.

10. Find the measure of the side of a rectangle if the length is 7 cm longer then the with and perimeter, 54 cm.

11. Jack can finish in job alone in 18 days, while Mario can finish in 10 days. How many days does it take for them to finish this job if they work together?

12. Jack can paint a wall in 10 hours, while his brother can paint it in 12 hours. How many hours does it take to finish it if they paint together?

13. Nine equivalent workers complete a job in 12 days when they work together. How many days does it take for 4 workers to complete this job?

14. A father is 25 years old, while his son is 5 years old. What will the ratio of their ages be in another 5 years?

15. The arithmetic mean of A and B is 14, while the arithmetic mean of C and D is 22. What is the arithmetic mean of A, B, C and D?

16. The arithmetic mean of three consecutives even numbers are 12. If the greatest one is 14, find the arithmetic mean of the other two numbers.

TEST-14

1.   2+4+6+...+38=?

2.   14+16+18+...+48=?

3. $5 \div 5 \div 5 \div 5$=?

4.   AA+BB=99, Find max value of AxB?

5.   AAA+BBB+CCC=888, Find value of (A+B+C) =?

6.   The product of two positive consecutive odd integer is 99. find the average of these integers?

7.   Find the sum of the first seven prime numbers?

8.   The product of two prime numbers is 39. Find the positive difference between these two numbers.?

9. Find the number of divisors of 80 ?

10. Find the number of divisors of 400.?

11. There are 360 students in a school. If 7/12 of the students are girls. Find the ratio of boys to girls.

12. Evaluate: $5005\frac{1}{5} - 2002\frac{1}{2}$

13. If the first day of a year is Wednesday, what day is the 206th day?

14. Find the arithmetic mean of the numbers 8, 6, 10 and 24

15. Mehmet can finish a job alone in 14 days, while Jack can finish in 16 days. In how many days can they complete this job if they work together?

16. Ahmet and Jack can finish a job in 15 hours when they work together, while Jack can finish it alone in 20 hours. How many hours does it take for Ahmet to finish this job alone?

TEST-15

1.   43.27+ 1.72+ 0.875

2.   A taxi driver charges \$12.74 plus \$3.84 per mile. Find the total cost of a-12 miles trip.

3.   Jessica earns \$7.29 per hour and \$8.94 for each hour over 38 hours per week. Find her earnings if she worked 46 hours last week?

4.   91/2 + 81/4 + 71/5 =?

5.   Teacher said 2/5 of students did not pass math last test. exam. if 36 students passed the exam. How many students are in class?

6.   If the scale on a map reads 3 inches = 69 miles, how many miles are there between two cities whose distance on a map is 9.5 inches.

7.   1/2 ÷ 1/2 ÷1/2 ÷ 1/2 =?

8.   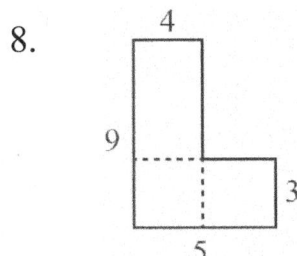   Area=
       Perimeter=

9. A number the sum of $\frac{1}{3}$ and $\frac{1}{5}$ is 40. What is the number square?

10. Find the volume of a cube witch length is 3.2 cm

11. Find the base area of a cube with length 8 cm

12. Find the lateral volume of a rectangular prism with side of length 3,4 and 5 cm.

13. ABC is a triangle. $\angle A=60$, $\angle B=70$. Find relationship to angle sides in ABC.

14. ABC is triangle AB=7 cm, AC=9 cm and BC= 11 cm. Find relationship the angles in ABC?

15. $A \div B = 5 \div 7$, Find $\frac{3A+1}{3B-1}$

16. If the first day of a year is Friday, what day is the 184th day?

TEST-16

1. 18+20 +22+...+44=?

2. Calculate the sum of 1/5 of 55 and 5.

3. Evaluate 21.20 + 3.75 =?

4. What digit in the ten thousand's place in 247,675,083.

5. Multiply 10.02 x 18.748 and round to the thousand this place.

6. What is the nearest integer to 12.65?

7. Compute the prime factorization of 60.

8. What is 0.34 written as a fraction reduced to lowest form.

9.   △ → 7   □ → 8   ⬡ → ?

10.   1+3+5+7+-----+55=?

11.   12+22+ 32 +------+102 =?

12.   Murat can paint a house in seven hours. Jack can paint the house in 12 hours. How long would it take to paint the house if they all work together?

13.   Find the product of the numerals of the greatest 2-digit prime number.

14.   Find the volume of the cube whose edge is equal to the smallest 2-digit prime number.

15.   Find the area of the square whose edge is equal to the sum of the first three prime numbers.

16.   The legs of a right triangle are proportional to 3 and 4. If the sum of the legs is 14cm, what is the perimeter equal to in cm?

TEST-17

1. 14.20+ 13.80 +----? = 64.94

2. What number is missing? 12,24, 36, ------- 60.

3. It is currently 1:45 pm. What time will it be in 55 minutes.

4. What number added to 96.74 is equal to 134.96.

5. Each book costs $8.74. How much does 9 books total cost?

6. What is the sum of the seven even whole numbers?

7. How many times does the digit "3" appear from 4 to 44.

8. How is the number 684 read in expanded form?

10. Look at the number pattern below. Which number comes next 3.2, 6.4, 9.6?

11. There are 5 cow, 4 lions and 6 eagles are at the zoo park. How many legs are there?

12. 144-12+12+12=?

13. A rectangle sides are in ratio 7:5. If the perimeter is 60 cm, find the area?

14. Jack can paint a home wall in 12 hours. It takes 18 hours for Mario to paint same wall alone. If they work together how long should it take to paint the wall?

15. Area=
    Perimeter=

16. In Angelina's rectangular garden, the ratio of the sides is 1 to 7. The sides are integers in terms of centimeter. Which one can be the garden perimeter?

TEST-18

1.   How many composite numbers are between 11 and 31.

2.   What is the remainder when 875 is divided by 3?

3.   What is the remainder when 177777 is by 3?

4.   Luis has 7 dozen books. How many books does he has?

5.   Find the product of 4, 5 and 7.

6.   Find the product of 23 and 32.

7.   Estimate 333+444+222

8.   Find next term in the arithmetic sequence 7, 13, 19, 25, 31, ----?

9.   Calculate: 0.25 - 1/4 =?

10.   Calculate: (2÷2÷2÷2) +2=?

11.   Calculate: 112+222 + 332+ =?

12.   Mario drove his car at 70 miles per hour for 6 hours. How for did he travel?

13.   Two dozen shoes cost $288.48. How much do three shoes cost?

14.   24.96 kg of watermelon costs $16.16. How much 3 kg of watermelon cost?

15.   The average of 5 numbers is 36.8. If 4 of the numbers are 17, 18, 19 and 24, what is the last number?

16.   Mario buys 9 dozen pencils. How many pencils does he have in all?

TEST-19

1.   4/14 + 10/14 + 1 =?

2.   9.75 + 9.75 + 9.75 =?

3.   0.48÷0.08=?

4.   (0.36) x (102) =?

5.   Find the ratio of 1/7 to 1/6 =?

6.   27 is to 3/5 of what number?

7.   Luis paid $64.32 for 8 books. How much did he pay for 5 books?

8.   The new geometry book was on sale for 4/7 of the original price. if the original price was $77. What was the sale price?

9.   A square's perimeter is 40cm. If the perimeter increases 20%. Find the new square's perimeter?

10. Find the area of new squares (above)?

11. 8417+6977+4321=?

12. 9874-6427=?

13. Raul drove his car 65 miles per hour for 6 hours. How far did he travel?

14. Mario can pedal his bicycle 9 miles per hour. At this rate, how long will it take him to ride 81 miles?

15. Jack traveled 475 km at an average speed of 25km/h. Find the time traveled in terms of hours.

16. Diego drove his car at 70 miles per hour for 8 hours. How far did he travel in terms of mile?

TEST-20

1. Find the average of 6, 8, 10 and 20?

2. 6, 8 and M average are 18. Find the value of the M?

3. Find the perimeter of an equilateral triangle on with side length is 7.36 cm.

4. The equilateral triangle perimeter is 21 3/4 cm. Find the one side length.

5. One complementary angle is 65°. Find the other angle.

6. One supplementary angle is 124°. Find the other angle.

7. 3 days = _____ hours.

8. 3600 minutes = _____ days.

9.   4 years = _____ days.

10.   12 raised to the second power is _____.

11.   8+64÷8+8 =?

12.   How many odd numbers less than 22.

Find the equivalent algebraic expression (Question 13-16)

13.   Five more than seven times a number.

14.   Five more than three times a number.

15.   9 decreased by the quotient of a number and 6.

16.   Six more than four times a number is 24.

TEST-21

1. How many composite numbers less than 19.

2. If 5A+5=75, What is 28- 2A=?

3. If 4a+4b= 36, Find 3a+3b value?

4. UTAH → 16,   KANSAS → 36
    TEXAS → 25,   OKLAHOMA →?

5. a + b=11, a × b=30, (a<b) find 2a+3b=?

6. Which small number are divisible by, 5, 7 and 9.

7. Which mall number are divisible by 3,5 and 7

8. How many rectangles is in figure?

    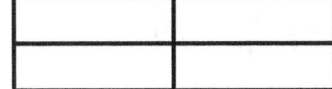

9. If SUN,SON = 987,967   Find (U + O)

10. 24 workers can complete a new home construction in 14 days. How many workers will be required to complete this task in 21 days?

11. 9 workers can paint a school's walls in 12 days. How many workers will be required to complete this school wall paint in 6 days?

12. The sum of the ages of 4 siblings is 86. Find the sum of their ages in 5 years.

13. A father is 36 years old, and his daughter is 14 years old. In how many years will the ratio of their ages be 2:1

14. 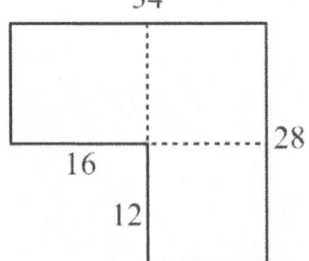   Area=
    Perimeter=

15. How many prime numbers are between 20 and 60?

16. Solve: $-\dfrac{6}{5}m - 4 = 14$

TEST-22

1.   ABC+ABC + ABC = 963, ABC is three-digit number.
     Find A+B+C =?

2.   How many triangles are in the figure.

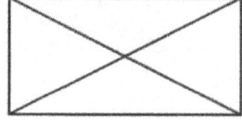

3.   Find the perimeter of the figure?

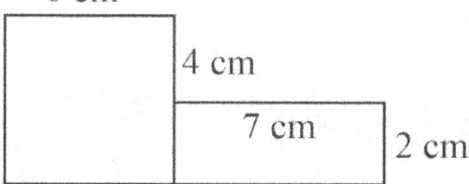

4.   7984+3227+ 148+54+5 =?

5.   127.647+24.83-14.73-6.784 =?

6.   1/12 ÷ 1/12 ÷ 1/12 ÷ 1/12 =?

7.   Find the sum of the positive even integer from 41 to 51.

8.   Murat buys five books. The price of the five books is $6.71, $7.64, $9.84, $19.84 and $39.99 How much does he spend in all?

9.   What is the smallest prime number bigger than 23.

10.   9764, what value does the circle
       number in the number above represent.

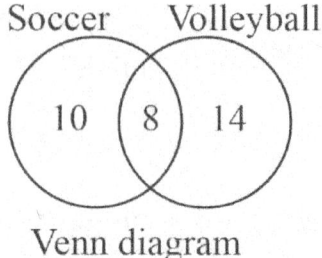

Soccer   Volleyball

10 (8) 14

Venn diagram

*   *The diagram shows a class after school club student registered.*

11.   What is the total number of students in class.

12.   How many students register both Club?

13.   The sum of the two even integers are 36. The larger of these two
       numbers is twice the smaller. Find the larger number.

14.   The difference in the squares of two consecutive odd integers is 24.
       Find the arithmetic mean of these numbers.

15.   The sum of the three consecutive odd numbers is k. Find the middle
       integer.

16.   The difference of 1/2 of an integer and 1/3 of the integer is 8. Find
       the integer.

TEST-23

1.   Calcúlate: 1x1+2x2+3x3+4x4=?

2.   If three book cost$21.27, how much will 7 book cost?

3.   Four times a number increased by 4 is equal to 24. Find the number?

4.   Calculate: (24 x 5/8) + (36 x 7/9)

5.   Ahmet can cut a lawn in 4 hours. His friend Jack can cut the same lawn in 7 hours. How long will it take them if they lawn at the same time.

6.   9.32+93.2+932=?

7.   Jock is 12 years old. His brother is 8 years old. Find the ratio age other 4 years.

8.   Find the mean of 4 , 6 , 8, 10 and 12=?

9.   Find the range of 5,21, 8, 32, 18=?

10.   Find the median of 22, 34, 28, 16, and 14=?

11.   How many miles will a car travel going 70 mph for 3.5 hours=?

12.   If the average of 40 number is 44, What is the sum of the numbers?

13.   The distance between two cities is 220 miles. First a car went one fifth of the road and then an additional 40 miles. How many miles left?

14.

Area=
Perimeter=

15.   Jack is 12 and Maria is 15 years old. Find the sum of their ages three years ago.

16.   Maria is 17 and Andre is 21 years old. Find the sum of their ages in 5 years.

TEST-24

1.   0.07+0.6+ 6.7+67=?

2.   (36-6) ÷ 6+6=?

3.   (9-1/9) + (9 +1/9) =?

4.   (3+1/3) + (3-1/3) =?

5.

      Area=
      Perimeter=

6.   The ratio of Murat's to Jack's age is 1:4.
      If the sum of their age is 25, what is their age difference.

7.   The sum of the ages of 4 siblings is 45. Calculate the sum of their ages in 6 years.

8.   What is the result when the numbers 475 and 624 are each rounded to the tens place and then added together.

9. If today is Friday, then what day of the week will it be in 160 days.

10. Find the sum of the positive divisors of 40.

11. ABCD is square perimeter of the square is 48cm2. Find they are ratio area to perimeter.

12. AB is two digit biggest prime number. Find value of 3A + 4B

13. It takes 12 hours for Ahmet to paint a wall. Jack can paint the same wall in 8 hours. How long would it take to paint the wall if they worked together?

14. Jack can paint a wall in 14 hours. Mehmet can paint the same wall in 12 hours. How long would it take to paint the wall if they worked together?

15. It takes 9 hours for Ahmet alone to paint a wall. Ahmet and Jack can paint the same wall together in 6 hours. How long would it take for Jack alone to paint the wall?

16. A pigeon can eat a box of food in 40 days. A hen can eat the same amount of food in 24 days. How many days would it take for the pigeon and hen together to eat a box of food?

TEST-25

1. HOME, SAME ↔ 1234, 5634 find the H+O+S+A=?

2. What is 1/9 of the largest two-digit positive number

3. The product of two is equal to consecutive odd integer 15. What is the sum of two integer?

4. The remainder of 4327 ÷5 is _____.

5. The remainder of (32 + 52) ÷ 4 is _____.

6. The product of 32 and 42 is _____.

7. 8837649, which digit is in the thousands place

8. 44+55+66+77+88=?

9. $14\frac{1}{2} + 13\frac{1}{5} + 15\frac{1}{4} = ?$

10. Compare $\frac{7}{11}$ and $\frac{6}{10} = ?$

11. Compare $\frac{7}{11}$ and $\frac{6}{10} = ?$

12. Compare $\frac{7}{11}$ and $\frac{6}{10} = ?$

13. $A \div B = 5 \div 7$, Find $\frac{3A+1}{3B-1}$

14. $A \div B = 2 \div 9$, Find $\frac{4A+4}{4B-4}$

15. $2 \times 3 \times 5 \times 6 \times 0 \times 10 = ?$

16. ABCD and EFLK are square . AB=10 cm, EF=6 cm. Find the shaded area?

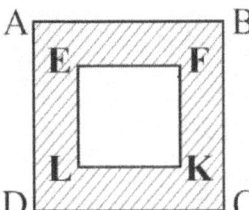

TEST-26

1. Find of the GCF of 36 and 60.

2. The GCF of the expression 9m and 3m is...

3. The sum of two integers is -3. The product of two integer is -54. What are two integers?

4. Two rational numbers have a sum of 22 and difference of 2. What are the largest number?

5. The sum of two integers is 16. The product of two integers is 60. What are two integers?

6. Which number is between $\frac{1}{4}$ and $\frac{2}{5}$?

7. Which number is between $\frac{7}{3}$ and $\frac{3}{4}$?

8. What is the surface area of a cube with a side of length 4 inch?

9.   What is the surface area of a cube with a side of length 10 cm?

10.   What is the surface area of a cube with a side of length 7 cm?

11.   What is the surface area of a rectangular solid that has a length of 6 cm, width of 4 cm and height of 8 cm.

12.   The volume of the rectangular prism is 60 cm^3. The height is 3 cm and the width is 4 cm. What is the length?

13.   In a survey of 300 students, 90 said liked the ice-cream. What percent is this?

14.   A map uses a scale 1 inch: 21 miles. What distance is represented by $7\frac{1}{2}$ inches on the map.

15.   If the train speed is 42 miles per hour and the train travels for 5 hours. How far will the train travel?

16.   Two angles in a triangle measure 42° and 64°. What is the measure of the third angle?

TEST-27

1. After school math club has 24 students. 11 of the students are boys. What is the ratio of girls to the total number of students?

2. Ahmet has 8 Algebra and 6 geometry books. What is the ratio of the number of geometry book to the total number of the book.

3. Book store sold 36 Algebra and 27 geometry books. What is the ratio of total number of geometries sold to the number of the Algebra.

4. At the bookstore, there are 42 Turkish book and 27 Germany books. What is the ratio of Turkish book to Germany books.

5. Jack spent $45 for a book. How much does sack pay for 4 books?

6. Two dozen pencils cost $12.How much does cost 4 pencil?

7. Ahmet walked 24 miles in 6 hours. At that rate, how many miles can Ahmet walk in 4 hours.

8. Ahmet earned $96.00 and 6 hours. How much Ahmet earn in $18\frac{1}{2}$ hours.

9.   Ahmet earned $84 and 7 hours. How much Ahmet earn in $6\frac{1}{2}$ hours.

10.   Math book that original cost $84. In on sale for 10% off. How much will a costume save.

11.   What decimal in simplest form are equivalent to 44%.

12.   What fraction is equivalent to 0.68.

13.   Jack answered $\frac{7}{8}$ of the question on his test correctly. What percent he did not answer?

14.   The rectangular garden has a length of 3x+4 and width of 2x+2. What expression represents the perimeter of the garden.

15.   A square pool has a length of 2x+3. Find the perimeter.

16.   Jack bought $2\frac{2}{3}$ pounds of watermelon for $ 1.30 per pound. How much did he

TEST-28

1. Is it possible for the side length 7 cm, 9 cm and 10 cm to form a triangle.

2. Two sides of the triangle have measures of 8 cm and 10 cm. What is the perimeter of the triangle?

3. A triangle`s one angle is measures 64°, another angle measures 32°. What is the measure of the third angle of the triangle.

4. Find the area of a square that has sides 3 cm long.

5. Find the perimeter of a square that has sides (2x+6) long.

6. A rectangular prism has a volume of 420 cm³/ Its height is 7 cm. What is its base are?

7. Two sides of a triangle have measures of 8 cm and 10 cm. What could be the length of the third side?

8. ABC is triangle. A=65°,B=75° Compare triangle side?

9. ABC is a triangle. AB=7cm, BC=9cm, AC=13cm. Compare triangle angles.

10. 12% of the eleventh grade at high school ride their own car. What fraction of the eleventh grade do not use own car.

11. Jack has $32.60. He spent $\frac{2}{5}$ of it for books. How much did the books cost?

12. If 12 math books are sold for $17.89 each. What are the total sales for the books.

13. Jack spends $48.24 on 4 math book that each cost the same amount. What is the price of each book.

14. What is $\frac{9}{20}$ written as a percent?

15. What is 96% written as a fraction in simplest form?

16. Jack has owned 60 books. 20% are math books. How many math books does he own.

TEST-29

1. 8 is 4% of what number?

2. 24 is 6% of what number?

3. The length of a rectangle is 6 cm more than the width. The perimeter is 32 cm .find the area of rectangle?

4. The length of a rectangle is 11 cm less than the width. The perimeter is 50 cm. Find the area.

5. The ratio of the sides of a parallelogram angle is 2:3. If the perimeter of the parallelogram is 60 cm, what is the length of the smallest side?

6. The sum of the squares of two consecutive even integers is 100. Find the two integers.

7. The sum of the squares of two consecutive even integers is 244. Find the numbers averages.

8. Add $\frac{2}{3}$ to the product of 3 and $\frac{1}{5}$

9. Add $-\dfrac{2}{5}$ to the product of -10 and $\dfrac{7}{25}$

10. Add $-\dfrac{2}{9}$ to the product of -7 and $\dfrac{5}{21}$.

11. The sum of three consecutive odd integers is 129. What is the largest of them.

12. What is 17 divisible by?

13. Twice a number less 3 is equal to four times the sum 5 and that number. What is the number?

14. The sum of three consecutive even integer is 72. What are they?

15. One even number is four times another. Their product is 64. Find the small number.

16. The similar triangles have side in the ratio $\dfrac{7}{9}$. What is the ratio of the areas of the triangles?

TEST-30

1. $5\frac{1}{2}$ gallon paint covers 300 square feet. How many square feet will 9 $\frac{3}{5}$ gallons of point cover.

2. The measures of the three angles of a triangle are $(2x)°$, $(4x)°$ and $(6x)°$. What is the great angle?

3. Write the number 3.4×10^4 in standard form.

4. Write $3\frac{3}{8}$ as a decimal form.

5. The ratio of the angles in parallelogram angle is 1:3. What is the measure of the smallest angle.

6. A square garden has an perimeter of 100 cm. Find the area of garden.

7. A computer repairmen charges $45 plus $30 per hour. Write equation for this situation.

8. A rectangle is 24 cm long. The perimeter is 84 cm. find the area of rectangle.

9. Two legs of a right triangle each measure (2x+4) cm. Find the area of triangle.

10. Jack has $3000 a saving account that pays 3% simple interest. How much interest is earned on this money in the first year?

11. A cube has a total surface area of 600 square cm. What is the volume of cube?

12. Jack ate dinner at a restaurant. The total bill was $84. The tip was 12% What was the amount at the tip?

13. The ratio of girls to boys in a school is $\frac{7}{9}$. There are 320 students at school. How many boys are this school?

14. Math book original price is $60. Jack bought this book for $45. By what percentage is the price of book reduced.

15. The price of a geometry book that regularly sells for $44 is reduced to $33. By what percentage is the price reduced.

16. Jack has 288 books. $\frac{3}{8}$ of the books are math books. How many of the books are math?

TEST-31

1. Algebraic expressions.
    $24xyz+15x^2y+18yx$ What factor is common to all three term.

2. $6x^2y+7xy+xym$ Find the numerical coefficients of each term.

3. Find the perimeter of a rectangle with sides measuring 4.3 and 12.7 cm.

4. Seven less than 4 times a number is 21. Find the number.

5. Nine less than 5 times a number is 26. Find the number.

6. Jack has a test store of 94, 82 and 81 points. What must he score on the fourth exam to have an average score of at least 88 points.

7. A $40 math book cost $44.20 with sale tax. Find the tax rate.

8. The rectangle is 4 times as long as it is wide and its perimeter is 100 cm. Find its area.

9. One number is 4 times another and their sum is 35. Find the product numbers.

10. If a truck can travel 81 mile on 3 gallon of gas, how much gas will it need to travel 324 miles?

11. A and B being positive integers if A:B=9:7, what is the minimum value of A+B.

12. The ratio of two number is $\frac{7}{9}$. If the smaller number is 28. Find the greater number.

13. What is the solution of 4(3x+2)=x+12+5x?

14. 20 less than $\frac{2}{5}$ of a number is 20. What is the number?

15. The sum of $\frac{2}{3}$ of a number and $\frac{1}{5}$ of the same number is 39. Find the number.

16. Ahmet`s age is m and Jack`s is n years old. What will the sum of their ages be in 3t years?

TEST-32

1. Six algebra and three geometry books cost $198, while two algebra and five geometry books cost $138. Find the one algebra and geometry book cost?

2. If the sum of two numbers is 32 and the difference is 2. Find the larger number square?

3. If the sum of two numbers is 20 and the difference is 4. Find the product numbers.

4. If the product of two consecutive even number is 168. Find the smaller number square.

5. The square of a number plus 6 is 55. Find the number cube.

6. A square of a number plus half its number is equal to 68. Find the number cube.

7. The cube of a number plus 6 is equal to 70. Find the number square.

8. If the sum of two even numbers is 26 and the product of two numbers is 168. Find the number ratio.

9. One number is 5 times another number if their sum is 30. Find the difference number.

10. One number is 5 times another number if their sum is 26. Find the difference number.

11. Find the sales tax on a computer cost $300 if the rate is 4%.

12. New math book originally price is $60? And was discount 22% for a sale, what was the discount price

13. If 6 geometry books cost $18.30. How much will a geometry books cost.

14. One odd number is 3 times of another. Their sum is 12. Find the product of them.

15. Jack has $60 in a saving account. He will start adding $8 to his account each day. Write equation in his account after x days.

16. Which side lengths turn a right triangle.
    I. 2, 4, 6   II. 6, 8, 10
    III. 3, 5, 7   IV. 1, 3, 5

TEST-33

1. Find the total cost of 4 pairs of algebra books at $48.72 each and 7 pairs of geometry books at $21.35 each.

2. Jack bought 4 pairs of chemistry books at $24.72 each. If he paid for them with a $100 bill, how much change did he receive?

3. A rental car driver charges $12 plus $3.72 per mile to find the total cost of a 11 miles trip.

4. Jack earns $72000 a year, what is the Jack`s 3 months' salary.

5. School principle said ¾ of student enrolled for next school year. If the 800 students study school. How many students did not enroll?

6. What percent of 35 is 25

7. 24 is what percent of 40

8. New school has 900 students if 60% of them are girls. How many of them student are boys?

9. New math book cost $28 and was discounted $7. What was the discount rate?

10. If tile sells for $12.18 per square. How much will it cost to cover a kitchen that measures 24.12 square.

11. Jack earns $7224 in 4 months; how much will he earn in $2\frac{1}{2}$ years

12. Find the Celsius temperature when the Fahrenheit temperature is 95. (use C= $\frac{5}{9}$ (F – 32)

13. University has 775 students if there are 75 more boys than girls. Find the ratio of boys to girls.

14. Find the interest on a loan of $6000 at 3 % for 7 years.

15. If Jack saves $82 for 2 month. How much will Jack have saved in 3 $\frac{1}{2}$ year.

16. Find the range of : 6,18,24,3,42

TEST-34

1. A right triangle has legs that measure 18 cm and 24 cm. What is perimeter of triangle.

2. $\angle A$ and $\angle B$ are vertical angles. If the measure of $\angle B$ is 64, find the measure of $\angle A$.

3. $\angle M$ and $\angle N$ are supplementary angles. If the measure of $\angle M$ is 68°. Find the measure of $\angle N$.

4. $\angle A$ and $\angle B$ are complementary angles. If the measure of $\angle A$ is 21°. Find the measure of $\angle B$.

5. $\angle M$ and $\angle N$ are complementary angles. If the $\angle M$ is 2x and $\angle N=$ 3x, Find $\angle M$

6. $\angle M$ and $\angle N$ are supplementary angles. If the $\angle M$ is 3x+2 and $\angle N=5x-8$. Find $\angle M$.

7. If the measure of an angle is 15°, find the measure of its supplement.

8. If the measure of an angle is 42°, find the measure of its complement.

9. ∠A and ∠B are vertical angles. If ∠A= 4x+14, ∠B=5x–12. Find value of x?

10. Find the midpoint of $\overline{KL}$ if K(4,8) and B(-2,-12).

11. If M is midpoint of $\overline{AB}$, find the coordinates of B if A(12,-4) and M(8,6).

12. The cost of tile is $0.36 per square cm. How much will it cost to the square 25 cm by 25 cm.

13. Jack went paint front garden wall. Front garden wall that is 15 m long and12 meter wide. What is the area of the painting?

14. A triangle has a base length of 12 cm and a height of 9 cm. What is the ratio area to perimeter.

15. Two side of a triangle measure 12 cm and 8 cm. Write inequality show to possible lengths for the third side x.

16. ABC is triangle. AB=8, AC=10 and BC=12 cm Write relationship the angles in ABC.

TEST-35

1. The ratio of two complementary angles is 1:2. Find the greatest measure of greatest angle.

2. The ratio of two supplementary angles is 1:5. Find the difference angle.

3. The ratio of the measure of the angle in a triangle is 2:4:6. Find the smallest angle.

4. The ratio of the measures of the sides of rectangle is 7:9. Find the minimum perimeter of rectangle.

5. Solve: $\dfrac{7}{a+3} = \dfrac{4}{a-4}$.

6. The ratio of the measures of the three angles in triangle is 1:2:3. Big angle is 90° Find the small angle.

7. ABC is triangle AB=AC, ∠A=70°, find the ∠C.

8. ABC is triangle AB=AC, ∠B=72°, find the ∠A.

9.   The ratio of the measures of the sides of a triangle is 2:3:5. Big side is 15 cm. Find the perimeter of the triangle.

10.   The ratio of the measures of the 3 sides of a triangle is 1:3:5. Small side is 2x+2. Find the perimeter of the triangle.

11.   The ratio of the measures of the parallelogram angle is 2:3. Find the small angle.

TEST ANSWERS

	1	2	3	4	5	6	7	8
TEST 1	131	28,35	2:55PM	1,2,3,4,6,8,12,24	26	1,2,3,6	63	76
	9	10	11	12	13	14	15	16
	26	53	20	14	48	2304	64	65/88
TEST 2	1	2	3	4	5	6	7	8
	5	798.77	88	35	$3.60	8:26pm	4588.05	58.46
	9	10	11	12	13	14	15	16
	144	72	78.67	15	$391.80	7to8	168	232/93
TEST 3	1	2	3	4	5	6	7	8
	6	>	2032	687	42	27	14	6
	9	10	11	12	13	14	15	16
	33	5	BC>AC>AB	30	41/42	17	37.70%	195/184
TEST 4	1	2	3	4	5	6	7	8
	900+70+5	769	20	38	3to8	144	27	124
	9	10	11	12	13	14	15	16
	110	24	90	12	$9.95	$38.24	240	35/6
TEST 5	1	2	3	4	5	6	7	8
	264	8	P=24,A=30	1175	500	7	102/5	23
	9	10	11	12	13	14	15	16
	10	87.163	P=70,A=152	136	20	66	21	6.8
TEST 6	1	2	3	4	5	6	7	8
	30/8	21	Even	0	3	17	16	20
	9	10	11	12	13	14	15	16
	350	84	46	18	P=32, A=44	32	-38/7	<
TEST 7	1	2	3	4	5	6	7	8
	11/25	855	33	16672	25	87.45	88	6
	9	10	11	12	13	14	15	16
	34.56	770	2793	77	P=28,A=32	1	150	9.7,14,11,11
TEST 8	1	2	3	4	5	6	7	8
	1728	1/3	17	24	2,17,34,68	12	14	32.4
	9	10	11	12	13	14	15	16
	96	7	26.31	10	24	10	10.8	4x+22

	1	2	3	4	5	6	7	8
TEST 9	a,b,c	c,b,a,d	c,a,c,d	d,a,b,c	y=4x	y=x+3	y=x+6	y=2x
	9	10	11	12	13	14	15	16
	y=x+9	11	78	P=30, A=40	60	19	175/144	19
TEST 10	1	2	3	4	5	6	7	8
	4,16,28	27	16	x<7	y>20	x<20	x<9	-8/3
	9	10	11	12	13	14	15	16
	y/4<10	70	16	23	18	12	impossible	14x+14
TEST 11	1	2	3	4	5	6	7	8
	324	3.2	100	54	150	15	342	100
	9	10	11	12	13	14	15	16
	1,2,4,8,22,44,88	284	500	30	77/60	199.728	333	84-38
TEST 12	1	2	3	4	5	6	7	8
	16.7	995.892	9	6	12	278.67	26	1,2,4,14,28
	9	10	11	12	13	14	15	16
	1,2,4,8	6	30	8	x+y+2t	2x-t	12	4
TEST 13	1	2	3	4	5	6	7	8
	100	33%	26.8	22,9.75,9,12	72	8	18to8	18
	9	10	11	12	13	14	15	16
	7,35	10,17	45/7	60/11	27	3	18	11
TEST 14	1	2	3	4	5	6	7	8
	380	522	1 TO 25	18	8	10	58	10
	9	10	11	12	13	14	15	16
	10	15	5to7	3003	Saturday	12	112/15	60
TEST 15	1	2	3	4	5	6	7	8
	45.865	$58.82	$384.54	$79.95	$60.00	$218.50	$4.00	P=28,A=39
	9	10	11	12	13	14	15	16
	$75.00	$32.77	$512.00	$70.00	AB>BC>AC	A>B>C	4/5	Sunday
TEST 16	1	2	3	4	5	6	7	8
	434	16	24.95	7	118.748	13	2x2x3x5	17/50
	9	10	11	12	13	14	15	16
	11	784	3162	84/19	63	1331	100	24

	1	2	3	4	5	6	7	8
TEST 17	36.94	48	2:40pm	38.22	$78.66	56	13	600+80+4
	9	10	11	12	13	14	15	16
	12.8	48	156	218.75	36/5	20	$78.66	
TEST 18	1	2	3	4	5	6	7	8
	14	2	1	84	140	736	900	37
	9	10	11	12	13	14	15	16
	0	2.25	666	420 miles	36.06	$1.94	106	108
TEST 19	1	2	3	4	5	6	7	8
	2	29.25	6	36.72	6to7	45	$40.20	44
	9	10	11	12	13	14	15	16
	48	144	19715	3447	390	9	19	560 miles
TEST 20	1	2	3	4	5	6	7	8
	11	40	22.08	7.25	25	56	72	2.5
	9	10	11	12	13	14	15	16
	1460	144	24	11	7x+5	3x+5	9-n/9	6+4x=24
TEST 21	1	2	3	4	5	6	7	8
	10	0	27	64	28	315	105	9
	9	10	11	12	13	14	15	16
	14	16	18	106	9	A=760, P=124	9	-15
TEST 22	1	2	3	4	5	6	7	8
	6	8	38	11418	130.963	144	230	$84.02
	9	10	11	12	13	14	15	16
	29	hundred	32	8	24	6	k/3	48
TEST 23	1	2	3	4	5	6	7	8
	30	$49.63	5	43	28to11	1025.52	4to3	8
	9	10	11	12	13	14	15	16
	27	22	245	1760	136	A=232,P=74	21	48
TEST 24	1	2	3	4	5	6	7	8
	73.77	5to2	8	18	A=108, P=72	15	69	1100
	9	10	11	12	13	14	15	16
	Thursday	90	3	55	4.8	84/13	18	15

	1	2	3	4	5	6	7	8
TEST 25	14	11	8	2	0	1344	7	330
	9	10	11	12	13	14	15	16
	42.95	>	<	>	4/5	3to8	0	64
TEST 26	1	2	3	4	5	6	7	8
	12	3m	9,6	12	10,6	3to10	2	96
	9	10	11	12	13	14	15	16
	600	294	208	5	30%	157.5 miles	210	74
TEST 27	1	2	3	4	5	6	7	8
	13:24	6:14	27/36	14/9	$180	$2	16miles	$296
	9	10	11	12	13	14	15	16
	$78	$8.40	11/25	17/25	12.50%	10x+12	8x+12	$3.80
TEST 28	1	2	3	4	5	6	7	8
	Yes	33cm	84	9	8x+24	60	15cm	AC>BC>AB
	9	10	11	12	13	14	15	16
	B>A>C	22/25	$13.04	$214.68	$12.06	45%	24/25	12
TEST 29	1	2	3	4	5	6	7	8
	200	400	55	136	4	6,8	11	19/15
	9	10	11	12	13	14	15	16
	-3.2	17/9	45	1,17	3	22,24,26	4	49/81
TEST 30	1	2	3	4	5	6	7	8
	585	90	34000	3.375	45	625	y=30x+45	432
	9	10	11	12	13	14	15	16
	2(x+2)(x+2)	$90	1000	$10.08	180	25%	25%	108
TEST 31	1	2	3	4	5	6	7	8
	3xy	6,7,1	34	4	7	95	1.05%	400
	9	10	11	12	13	14	15	16
	196	12	16	36	2/3	100	45	n+m+6t
TEST 32	1	2	3	4	5	6	7	8
	$24,$18	15	96	144	343	512	16	6to7
	9	10	11	12	13	14	15	16
	32	36	12	$13.20	$6.05	27	y=60+8x	2

	1	2	3	4	5	6	7	8
TEST 33	$344.33	41.12	$.52.92	$18,000	200	140%	60%	360
	9	10	11	12	13	14	15	16
	25%	$293.79	$50,568	35	17/14	$1,260	$924	$39
	1	2	3	4	5	6	7	8
TEST 34	72cm	64	116	69	36	60	165	48
	9	10	11	12	13	14	15	16
	26	(1,-2)	(4,16)	$175	180	3to2	4<x<20	A>B>C
	1	2	3	4	5	6	7	8
TEST 35	60	120	30	19	40/3	30	55	36
	9	10	11					
	30	18X+16	72					

About the Author

Tayyip Oral - Math Teacher, Author, and Education Expert

Tayyip Oral is a distinguished mathematician and education innovator, born Van city, in Turkiye. With a rich academic background, he holds a Master's in Business Administration and a Master's in Education Leadership from Qafqaz University and North American University, respectively. Since 1998, Mr. Oral has been imparting his mathematical wisdom, teaching at various learning centers and high schools.

Renowned for his contributions to mathematics education, Mr. Oral is the brain behind the successful 555 math book series. This series showcases an array of math books designed to cater to different learning needs. His expertise extends beyond traditional publishing; he has authored over 20 books and developed 8 educational math apps. His works include SAT Math, ACT Math, Geometry, Math Counts, and Math IQ books.

Mr. Oral's passion for mathematics and education is evident in his hands-on approach to teaching and curriculum development. His innovative methods and dedication have made a significant impact on students and educators alike. He currently resides in Houston, Texas, where he continues to inspire and educate future generations of mathematicians.

For more information about his work and contributions, Mr. Oral can be reached at **tayyiporal@gmail.com**, and his projects can be explored at **www.555math.com**

ACKNOWLEDGEMENT

I want very special thanks to my students.

1. Ahmet Mahir Inalhan   2. Nihal Gumus

3. Muharrem Kendirci   4. Daniel Gonzalez

5. Megh Patel   6. Ethan Patel

7. Alexander Gleason   8. Lincoln Vu

9. Ulugbek Abdurraimov   10. Nathan Chiu

11. Dmitry Lounev   12. Callan Joshua Nicole

13. Nozima Agzamova   14. Aaron Moniaga

15. Martin Bate   16. Edgar Morales

17. Andrew Hieule Nguyen   18. Reyash Maheshwari.

19. Esma Alizade   20. Kerem sarioglu

For your effort and feedback.

Tayyip Oral
555 MATH BOOK SERIES AUTHOR

1/15/2024

Books by Tayyip Oral

1. Sheryl Knight, Mesut Kizil, Tayyip Oral, ACCUPLACER MATH PREP, (1092 Questions with Answers), 2018

2. Sheryl Knight, Mesut Kizil, Tayyip Oral, TSI MATH Texas Success Initiative, (1092 Questions with Answers), 2017

3. Tayyip Oral, Osman Kucuk, Hasan Tursu, Geometry for SAT & ACT (555 Questions with Answers), 2017

4. Tayyip Oral, Ferhad Kirac, Bekir Inalhan, Algebra for The New SAT, Level – 1, (1111 Questions with Answers), 2017

5. Sheryl Knight, Tayyip Oral, Servet Oksuz, Algebra for ACT, Level – 1, (1080 Questions with Answers), 2017

6. Kristin Alexander, Tayyip Oral, Sait Yanmis, 555 Gifted and Talented, Question Sets for the Mathematically Gifted Middle Grade Scholar (1111 Questions with Answers), 2017

7. Steve Warner, Tayyip Oral, Sait Yanmis, 1000 Logic&Reasoning Questions for Gifted and Talented Elementary School Students, 2017

8. Tayyip Oral, 555 ACT Math, 1110 Questions with Solutions, 2017

9. Tayyip Oral, 555 Math IQ for Elementary School Students (1270 Questions with Answers), Second Edition, 2017

10. Tayyip Oral, Ersin Demirci, 555 SAT Math, 2016

11. Tayyip Oral, Sevket Oral, 555 ACT Math - II, 555 Questions with Answers, 2016

12. Tayyip Oral, 555 Geometry (555 Questions with Solutions), 2016

13. Tayyip Oral, Dr. Steve Warner. 555 Math IQ Questions for Middle School Students: Improve Your Critical Thinking with 555 Questions and Answer, 2015

14. Tayyip Oral, Dr. Steve Warner, Serife Oral, Algebra Handbook for Gifted Middle School Students, 2015

15. Tayyip Oral, Geometry Formula Handbook, 2015

16. Tayyip Oral, Dr. Steve Warner, Serife Oral, 555 Geometry Problems for High School Students: 135 Questions with Solutions, 2015

17. Tayyip Oral, Sevket Oral, 555 Math IQ questions for Elementary School Student, 2015

18. Tayyip Oral, 555 ACT Math, 555 Questions with Solutions, 2015

19. Tayyip Oral, Dr. Steve Warner, 555 Advanced math problems, 2015

20. Tayyip Oral, IQ Intelligence Questions for Middle and High School Students, 2014

21. T. Oral, E. Seyidzade, Araz publishing, Master's Degree Program Preparation (IQ), Cag Ogretim, Araz Courses, Baku, Azerbaijan, 2010.
    A master's degree program preparation text book for undergraduate students in Azerbaijan.

22. T. Oral, M. Aranli, F. Sadigov and N. Resullu, Resullu Publishing, Baku, Azerbaijan - 2012 (3.edition)
    A text book for job placement exam in Azerbaijan for undergraduate and post undergraduate students in Azerbaijan.

23. T. Oral and I. Hesenov, Algebra (Text book), Nurlar Printing and Publishing, Baku, Azerbaijan, 2001.
    A text book covering algebra concepts and questions with detailed explanations at high school level in Azerbaijan.

24. T.Oral, I.Hesenov, S.Maharramov, and J.Mikaylov, Geometry (Text book), Nurlar Printing and Publishing, Baku, Azerbaijan, 2002.
    A text book for high school students to prepare them for undergraduate education in Azerbaijan.

25. T. Oral, I. Hesenov, and S. Maharramov, Geometry Formulas (Text Book), Araz courses, Baku, Azerbaijan, 2003.
    A text book for high school students' university exam preparation in Azerbaijan.

26. T. Oral, I. Hesenov, and S. Maharramov, Algebra Formulas (Text Book), Araz courses, Baku, Azerbaijan, 2000
    A university exam preparation text book for high school students in Azerbaijan.

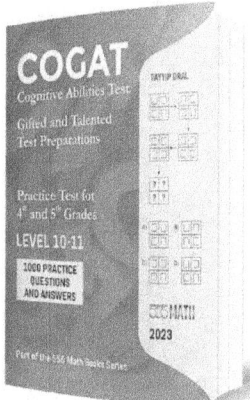

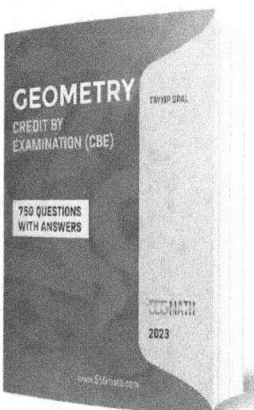

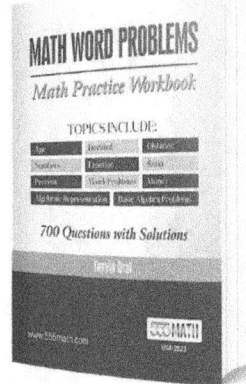